Glimpses of Wonder

Todd Duncan

Science Integration Institute
Beaverton, Oregon

Glimpses of Wonder

Published by
The Science Integration Institute
http://www.scienceintegration.org
info@scienceintegration.org

ISBN 978-0-9712624-2-3 (paperback)

To my father.

With gratitude for encouraging wonder, and especially for nights spent looking up at the stars together.

Contents

Introduction

*"The most beautiful and deepest experience...
is the sense of the mysterious... a something
that our mind cannot grasp and whose beauty
and sublimity reaches us only indirectly and
as a feeble reflection..."*
~ Albert Einstein

Profound mystery permeates each moment of our experience. But the mystery is often veiled beneath the details of everyday life. Narrow focus gives the illusion of understanding and control. We know how to brush our teeth, walk to the store and buy a loaf of bread, send music and videos invisibly through the air around the globe, unconsciously follow conventions set by our society, even land people safely on the Moon. We have to be shaken up a little to remember that from a broader perspective it's completely mysterious how any of this—including teeth, bread, social conventions, moons, and music—exists at all. We are part of something much greater than our individual selves whose essence we only dimly perceive.

Glimpses of this mystery at the heart of reality sometimes peek through the veil, echoing and reverberating throughout our being. The sublime surprise that anything exists at all floods us with a sense of wonder and awe. From this perspective everything is enchanted, the simplest and most mundane actions imbued with the mystery of existence itself. Such moments sometimes occur when we first wake up—perhaps because while sleeping we partially forget that we exist. So there's a brief instant as awareness returns when we reorient ourselves to this fact and are amazed by it once again.

It's worth cultivating such glimpses of wonder because not only are they portals into what is most real and fundamental about our existence, but they can be a deep source of inspiration, joy, and motivation. Consider the difference between your narrow frame of mind when someone cuts you off in traffic at the end of a hectic day, and the expansive feeling of connection while lying under the stars on a summer night wondering if someone out there is looking back at you.

Try recalling examples from your life. When is your awareness closed and narrowly focused on your immediate surroundings? When is your awareness open and expanded like a web whose threads stretch out to the most distant times and places in the universe?[1] What triggers the shift for you, from small web to large network of consciousness? Walking in the forest? Looking up at the stars? How about arguing with a friend or taking a test? When do you feel most connected to other people and the rest of the world around you, and when do you feel most isolated and alone? When do you feel inspired and when do you feel unmotivated?

Each essay in this book is meant as a portal to help transport you beyond the everyday veil and bring the ultimate mystery of existence into clearer view. When you can glimpse in your mind's eye the thread connecting even the most mundane or frustrating everyday experience (such as an illness, broken heart, boring job, or being stuck in traffic with obnoxious drivers) to the mystery at the heart of reality, your world is transformed and you feel directly the "rapture of being alive."[2]

I hope you'll be able to pick up this book when you're feeling disconnected and narrow, read for a few minutes, and come away with a renewed feeling that your life is a thread in the tapestry of something amazing beyond comprehension. And as that awareness seeps through you I hope its energy will help feed and sustain your enthusiasm for life, no matter what challenges and joys it throws your way.

Chapter 1

Cosmic Ancestry

Let your mind drift back over past stages of your life, both recent and distant. Can you recall the experience of eating breakfast today, the taste of your cereal, toast, yogurt, coffee or orange juice? Emerging from the mist of sleep into consciousness this morning? Cold air hitting your face as you walked out the door? First arriving at the place you now call home? Starting your current job, or your first job? Your first kiss? Your first day of school? A favorite story or song you fell in love with as a young child? The first time you recall being aware of anything at all? Pause to savor each memory, letting it become vivid and real in your mind's eye.

You are not the same person in the present moment as you were in any of those past moments. Your self has changed over time. Yet there is also a thread of continuity unifying your memories. Even the very distant ones that seem like a dream are connected by the single continuous thread that is your awareness.

Now imagine extending this same thread beyond your individual birth, to encompass the whole continuous line of your ancestry. Imagine incorporating, into the same chain of memories you can recall, the memories of your mother or father, and her mother or father, and so on. Continue this chain back through all the life forms that comprise the thread of your ancestry. Each step back makes the memory a fainter and fainter echo, even more tenuous than your first day of kindergarten, but a real memory nevertheless. Recall how you felt as a pre-historic human expressing yourself with language or harnessing fire for the first time. Remember your experience as a shrew-like creature perhaps fifty million years ago, feeling fear as it ran from a predator, but with no language to label the feeling. A worm recoiling in response to touch. Was there any accompanying awareness at all? Can you stretch your memory to recall the dawn of awareness itself?

What did that feel like? Did the first moment of awareness in the cosmos happen here on Earth, or perhaps somewhere far away across the universe?

Memory fails at earlier times but the thread continues: Your great-great-great...grandmother was a one-celled organism. And before that your "ancestors" existed as non-living forms: complex chemicals in a primordial soup on the early Earth and elements like carbon, calcium, and iron that were manufactured in the cores of ancient stars then blown off as clouds of gas and dust during the death stages of these previous stellar generations. We can see many examples of these clouds, containing material very similar to our own distant ancestors, in the breathtaking photos provided by the Hubble Space Telescope and other telescopes. Even simple hydrogen, formed in the very hot environment of our early universe 13.7 billion years ago, is one of your ancestors.

The pattern forming your identity as you read these words is a fossil record of the entire history of the cosmos. It expresses a much vaster pattern spanning billions of years and including Earth's ecosystem that nourishes and sustains you, nuclear reactions in the Sun's core from which light pours out to drive and sustain our ecosystem, ancient stars whose deaths produced elements that now make up your body, and earlier cosmic conditions that created the environment in which these stars could form. You are a moment of awareness within one strand of a vast cosmic tapestry. You are truly a child of the stars, and of the whole universe.

Chapter 2

Cosmic Connections, Distant & Nearby

While walking home one night I was feeling slightly detached from everything, for no obvious reason. But the sky was clear and dark despite nearby city lights so I paused to look up at the stars for a moment. My mood was transformed almost instantly.

First I was struck by the shock of direct awareness that reality is much bigger than me. Of the few dozen stars I could see, most are tens or hundreds of light years away.[3] So the light I see probably set out toward me before I was born.[4] I'm connected to those distant stars and whatever beings likely live among them. The light carrying our stories intersects in crisscrossing paths throughout the universe, weaving us together even if we never meet or even know of the other's existence. We share elements perhaps forged in the deaths of each other's home stars. The single continuous chain of ancestry discussed in chapter 1 is illuminating but not the whole story. Each thread is linked to many other threads, forever entwined like ideas whose exact origin is lost in the entangled memory of conversations with many different friends. We're interconnected in a vast cosmic tapestry.

As I scan my eyes across the sky they probably take in light from at least one star supporting intelligent life that we may never know about. Yet this life almost certainly exists, with feelings perhaps something like mine and also completely alien to my experience.

I kept walking toward home and spotted the neighbor's cat, Lucy, perched in front of her garage door. She looked at me intently so I sat down on the sidewalk. Immediately she ran over and started circling, rubbing against me, purring loudly, delighted at the attention. Or so I supposed from my biased anthropocentric perspective. I can't purr but internally I felt the same way, uplifted by such a simple connection.

Apparently recognizing there was affection to be had, another cat appeared out of the darkness and ran over, stretching out on its back next to me to be petted. We three enjoyed each other's company for awhile.

Once I was sufficiently covered in fur I got up to leave. I paused in my doorway and watched the two cats, apparently back to their regular nightly behavior now that I had vanished from their awareness. One stalked a shadow invisible to me. Lucy watched from a distance for a moment, then followed her friend. I felt I was watching alien life just as surely as if I'd been transported to one of those distant stars glanced at earlier. I could partially relate to the cats' actions and feelings and of course had shared a little moment of connection with them. Yet in many ways I had no idea what they felt, how they experienced the world, what drove their exploration.

In how many places and in how many forms throughout the universe are similar experiences playing out?

Chapter 3

Aliens Among Us?

Close your eyes for a moment and picture what lies in front of you during the rest of the day. What tasks await on your to do list? What phone calls need to be made, emails sent, projects or homework assignments completed, sinks cleaned, lunches made, dogs walked? What's your attitude toward them? Excitement, dread, boredom, acceptance, laziness?

Now imagine your experience when that commonplace state of awareness is pierced by confirmation that an intelligent civilization is orbiting the star 51 Pegasi. How would your view of the rest of the day change? What emotions would you feel in the instant you heard about it, and in the instant you became convinced it was true?

I would be dramatically transformed but it's difficult to pin down exactly why, since I already feel pretty confident that there is intelligent life out there somewhere. I'd feel a rekindled sense of wonder, awe, amazement. I'd want to know how they came to be, what they think about, what they feel, what they experience that is similar to my own experience and what parts of their awareness have no analogs for me. The atmosphere would be electric. A sense of magic would embrace the way I looked at the world—something akin to falling in love. Any mundane hazy feelings about life would vanish. All those tasks I face (some of which I would ignore now as I plunged into investigating this new discovery) would look totally different, embedded in a magical context. I would be transported into mystery, immersed in the reality of the basic questions of existence. How did life arise, and self-awareness? What do the existence of these things say about the meaning of existence itself?

Now consider this: to those beings orbiting 51 Pegasi (or wherever there may actually be intelligent civilizations), *our* existence here on Earth represents that magical, perception-changing discovery! The Pegasians might feel their world transformed by the knowledge of our existence. This is true even though we would of course be unchanged in the moment of their discovery. We would still be just who we are with all our abilities, faults, and difficulties. (And depending on how far away they are from us we might even be long gone at the moment our light reached them, bringing to their awareness the news that we once existed. See chapter 5 if you'd like to ponder this issue further.)

What are we to make of this fact? It means our sense of wonder is a matter of *perception*. The existence of aliens does not create the wonder. It only serves as a trigger, a portal, reminding us of a perspective that is available all the time. The triggers matter of course. We can't always turn that sense of awe and wonder on and off at will, but still it *is* a matter of perception. You can shift your frame of mind so everything you experience in daily life is like the discovery of aliens. You're fully justified in treating every interaction with every part of the world with that same sense of wonder and amazement!

Another way to look at it: there are "aliens" among us all the time if we only shift our frame of mind to take notice. *You* are an alien to beings on another planet, and on this planet. Dolphins, whales, chimps, cephalopods, your cat or dog, even other humans are in a very real sense aliens to you. (For that matter, don't you often feel even that you are an alien to *yourself*, with thoughts and feelings you can't completely explain or understand?) Look into their eyes for a moment and you know you are in the presence of another being with a view of the world all its own, alien from your perspective.

Whenever things seem a little mundane, close your eyes and try to recreate the feeling of wonder and magic you would experience if we discovered extraterrestrial life. Savor it, mull it over in your mind, play with how it feels in your body. Then recall once again that to those aliens you are the alien life. You do exist. So the wonder is yours. It is real. Go visit your neighbor and view them as if they were an alien with a totally new perspective to share with you. (Admit it, you've always suspected your neighbor was an alien!) You might even try transforming any judgments you feel toward people you interact with. What if you replaced those judgments with the curiosity you would experience while talking to a completely alien life form with something to teach you?

Chapter 4

The Mystery of Time

*"What, then, is time? If no one asks me, I
know what it is. If I wish to explain what it is
to one who asks me, I do not know."*
~ Saint Augustine of Hippo

*"The gods confound the person who first
found out how to distinguish hours. Confound
him, too, who in this place set up a sundial,
to hack and cut my days so wretchedly into
small portions!"*
~ Titus Maccius Plautus

Our familiarity with time usually disguises its
deep mystery. But the mystery lurks just beneath the surface, waiting patiently to become
visible whenever we pause to notice it. As physicist James Jeans wrote, "So little do we understand time that perhaps we ought to compare
the whole of time to the act of creation." Let's
consider a few observations to help peel back
the surface layers of our everyday familiarity
with time, revealing the glimpse of wonder
waiting beneath. Sometimes we need a nudge
to shake us out of our complacency so we can

be properly amazed by what's all around us. (Much as a fish might not notice the water it swims in until it's lifted out and then dropped back into the pond.)

As I write, I hear the faint tick-tick of my watch announcing the steady passage of time that dominates my life. For that matter, I don't really need a watch for this. Time is embedded in everything. Clocks just quantify the flow and make it more visible. My heart beats at regular intervals. The Sun rises and sets and the Moon moves through the sky from night to night against the backdrop of stars. An apple slice on the counter grows browner with each passing hour. A swing glides back and forth, back and forth. Start counting the number of swings and you have a crude clock. Make the swing very tiny and add an automatic mechanism for counting the oscillations, and you have a modern watch.

If we synchronize our watches and then you walk to the other side of town, your watch will read 10:37 at precisely the same moment mine does. An obvious fact we take for granted. But wait a minute. How does my watch know how fast to tick in the first place? Is it embedded in something, some property of space that provides instructions telling it when to tick each second or fraction of a second? And however it gets its instructions, how does *your* watch, in a *different* part of space way across town, know to get the *same* instructions?

In case you think I'm making a big deal out of nothing, consider this additional twist to the story: if your watch is moving rapidly relative to mine, the instructions our watches receive for how fast to tick will get out of sync. What's more, their discrepancy will follow a precisely defined rule that depends on our relative speeds! This rule is described by Albert Einstein's famous Special Theory of Relativity (parent of the well-known equation $E = mc^2$), now over a hundred years old and well-established by many experimental tests. Clocks (of any sort, including the internal ones that tell you how fast to age and when to sleep or feel hungry) in motion relative to each other do *not* tick at the same rate. If you are moving fast enough relative to your twin sister, you will age at different rates. Dramatically so, if your speeds get close to the speed of light. For example, at a relative speed of 99.9% of the speed of light (about 299,493 km/sec) a year for one twin is about 22.4 years for the other.

With this fact in mind, it's much harder to take the ticking rate of clocks for granted. The question returns with far greater force. "How do clocks know how fast to tick?" Something to consider the next time you're late for an appointment and feeling that panic in the pit of your stomach!

"What is time? The shadow of the dial, the striking of the clock, the running of the sand, day and night, summer and winter, months, years, centuries-these are but arbitrary and outward signs, the measures of Time, not Time itself. Time is the life of the soul."

~ Albert Einstein

Chapter 5

Build Your Own Time Machine

Stand in front of a mirror, about half a meter away, and carefully observe what you see. Wave to your time-traveling, past self. Believe it or not, you've just constructed a simple time machine! Every time you look at your reflection, you're peering about one three-hundred-millionth of a second into your past.

Let me explain. And in case you're not impressed with the limited temporal range of this simple time machine, I'll also tell you how to peer millions and even billions of years into the past!

Light travels so quickly that we often assume it takes no time at all to get from point A to point B. In truth it covers an impressive yet still finite 300,000 km each second. You can think of each bit of light as a messenger, carrying with it a snapshot of the scene it left behind.

But once in transit, the light is disconnected from the events it has recorded. For example, bits of light reflecting from your face soon after you were born traveled away from you at the speed of light, carrying the image of your face as an infant. If they traveled unimpeded, these bits of light would reach the star Proxima Centauri a little over four years later. A "Centaurian" looking at Earth with a powerful telescope, zoomed in on your face, would intercept those bits of light and literally see you as an infant. This is true even though, at the instant the Centaurian sees the light that is the image of you, you are no longer an infant. Back on Earth, you have transformed into a self-confident four-year-old while the light was on its long journey to Proxima Centauri. The same situation occurs when you look at yourself in the mirror. You just don't change much in a three-hundred-millionth of a second, so you don't notice the delay.

Your time machine becomes more powerful as you look at more distant objects. Consider the Whirlpool Galaxy (M51), about thirty-one million light years away from us. (You can view a nice photo of this galaxy at hubblesite.org, among many other places.) Quite likely among the hundred billion plus stars in the Whirlpool Galaxy, there is at least one with intelligent life on a planet orbiting that star. Much less likely, but still possible in our imagination, this civilization is so advanced that they have telescopes capable of zooming in to look at our tiny Earth. What would they see if they were curious to check up on you, focusing in on the exact spot where you are sitting under a tree reading this book? Give the question a little thought before you answer and then read on...

What they would *not* see is you intently reading a book. Instead, they would see whatever was happening in that spot *thirty-one million years ago*! And similarly, when you look at an image of their galaxy, you are seeing it as it *was* thirty-one million years ago. Your time machine has grown far beyond its small beginnings as your bathroom mirror!

It's worth pausing to reflect a bit on what this implies for both our connection to and isolation from the rest of the universe. The vastness of the star-filled cosmos lends weight to the belief that there must be other life, other civilizations even, somewhere out there. (See chapter 10 for more on this.) But that same vastness means we are likely very isolated from any civilizations that may exist. And not just isolated by distance, but by *time* as well. Civilizations are probably like fireflies flashing briefly in a huge dark field, disconnected by great distance and their short lifespans.

It helps to put this in perspective with an example. We humans have possessed the technological capability to send messages out into space for less than a hundred years. This means that only aliens within about a hundred light years could possibly have received any message from us. (Perhaps they are at this moment watching "I Love Lucy" reruns!)

Only a tiny fraction of the stars even in our own *Galaxy* fall within this range. Our Milky Way, the galaxy we reside within, is about a hundred *thousand* light years from edge to edge. Imagine drawing a sphere around our solar system with a radius of one hundred light years. This sphere represents the approximate region of the Milky Way our TV and radio signals have had time to reach. Now here's the punchline: That sphere (dot, really) is *way too small* to show up on a drawing of our Milky Way. (The diameter of the dot is about one five-hundredth of the diameter of the Milky Way, if you want the details.) Another way to think about it: Assuming you are less than a hundred years old, nobody outside of that dot could possibly know you were born yet. There has not yet been time for that joyous news to have reached them, even traveling at the speed of light! (If you are more than a hundred years old, congratulations! You get to draw a sphere slightly bigger than the one I've been talking about... and it still won't show up on a drawing of our Galaxy.)

You're probably feeling very tiny right now. The point is not to be depressed about how small you are, but to realize how vast is the cosmos in which you are an interconnected strand. Spend some time browsing through images at

Astronomy Picture of the Day or hubblesite. org or your own favorite site for astronomical images. Note the distances to the objects you look at. Some of them are billions of light years away. Savor the experience of peering billions of years into the past with your time machine.

Chapter 6

Transformations and Energy

Transformations pervade our lives. Like it or not, things change. The concept of *energy* pervades these transformations. We're familiar with the term in everyday conversation. We know we're supposed to conserve energy. We even come face to face with numerical amounts of it when we pay our power bill. But how well do we really understand what a "kilowatt-hour" *is* or how that number on our bill is connected to specific impacts on the environment? Energy is familiar and ordinary, yet it's also a subtle and mysterious concept which leads to deep awareness of our interconnection with the rest of the universe.

The modern, formalized definition of energy can seem abstract and unfamiliar, but the concept emerged as a unifying principle for commonly observed patterns in our direct experience. A few simple observations: You get tired when you climb a long flight of stairs. You feel hungry if you exercise for awhile without eating. A lamp won't shine if it isn't plugged in. Your car stops moving if it runs out of gas. You can be cold lying on a beach under cloud cover, and turn toasty warm when the sun comes out. You speed up as you coast downhill on a bicycle, and you slow down again if you try to coast back up the next hill.

All these experiences share the feature that a particular change to the world is accompanied by other corresponding changes. You can coast from the bottom of the hill to a higher point, but your speed will be slower at the new position. You can move from the bottom of the stairs to the top but not without feeling a little more tired. You can make your car move but not without using up gas. Feeling warmer while lying on the beach requires sunlight streaming down upon you.

These observations suggest a general principle: In order to change something, a "capacity" for change is always needed and this capacity must be *taken away* from one part of the world and *given* to the part you want to change. With a little refinement to turn it into a quantitative measure, this "capacity" is what we now call energy. One of the deepest discoveries in all of science is the law of *conservation of energy*. As its name implies, this law says that the total amount of energy in the universe is *conserved*— it never changes. This means you can take energy from one place and move it somewhere else or change it from one form into another (e.g. from energy in sunlight to energy in hot water). But if you add everything up, making sure nothing slipped away unnoticed, the total amount of energy stays the same. In other words, *the amount of energy lost in one place always equals the energy gained in another place.* Viewed from this perspective, energy provides a tangible, precise expression of our vague intuition that everything is interconnected. Energy is the common thread that links everything together. It's the constant in a world of change.

It's worth delving into the numbers a bit to reveal the richness hidden in this simple idea. We can associate a numerical amount of energy with any part of the world we care to pay attention to. There are very specific rules for calculating the amount. You've probably heard of some of them. Kinetic (motion) energy is calculated by multiplying the mass of the moving object (say a squirrel running along a power line) times the square of its speed and dividing by two, which gives the formula $1/2 \, mv^2$. Gravitational potential energy (stored in the squirrel's position high off the ground) is given by mgh—the mass of the squirrel times the acceleration of gravity (about 9.8 m/s^2 at Earth's surface) times the squirrel's height off the ground. Another familiar equation expresses the energy stored in the fact that an object has mass, $E = mc^2$, where c is the speed of light (about 300,000 km/s). This equation reveals that matter is a very concentrated form of energy. Because the speed of light is a large number, a small amount of mass stores a tremendous amount of energy, a fact which is made shockingly apparent in a nuclear explosion that releases some of this mass energy into other forms.

Other types of energy include heat, light, sound, and chemical bonds (such as those in food). In fact everything is energy in some form. All the action in the world is about the flow of energy from one form to another. Energy from sunlight gets stored in chemical bonds inside an acorn. A squirrel eats the acorn and then transforms the energy into gravitational potential energy as it climbs up to the power line. If the squirrel slips and falls, gravitational potential energy is converted into kinetic energy as the squirrel builds up speed while falling. I'll leave it as an exercise for the reader to list possible forms the energy might take next. Or if you'd like, you can imagine that just before reaching the ground, the squirrel encounters its anti-squirrel and the mass of both squirrels is transformed into a brilliant flash of light.

Thanks to the law of conservation of energy, we can pick any form we like as a standard of reference for measuring the amount of energy stored in any other form. For example, a **calorie** is defined as the amount of energy required to raise the temperature of 1 gram of water by 1 °C (starting at a standard reference temperature of 14.5°C). The familiar food **Calorie** (capital "C") is equal to a kilocalorie or *1,000* calories. Most people eat about 2,000 to 3,000 Calories per day—enough energy to heat 2,000 to 3,000 kilograms of water by 1 °C. It's helpful to know a few other units for reference:

joule – Another unit of energy, equal to the amount of work done in exerting a force of 1 newton through a distance of 1 meter. 1 calorie is 4.2 joules, so 1 food Calorie is equivalent to 4,200 joules.

watt – A unit of power, which measures the rate at which energy is transferred from one form to another. Power has units of energy divided by time; so just as energy and time can be measured in many different units, so can power. The watt is defined as 1 joule of energy transferred each second.

kilowatt-hour – Yet another unit of energy, which you have probably seen on your power bill. The odd unit ("1,000 watts times one hour") arises because we measure the rate of energy flow from the power company in the convenient unit of a kilowatt (1,000 watts or 1,000 joules transferred per second) and multiply it by the number of hours during which we draw power. So a kilowatt-hour is also (1,000 joules/second) x 3,600 seconds = 3.6 million joules. This is an odd but perfectly correct unit of energy. It's analogous to measuring distance in units of something like "miles per hour times seconds." Normally you would multiply miles per hour by the number of hours traveled, to get simply miles. But you could also multiply by another unit of time such as seconds, and it's still a valid distance.

It's also handy to have in mind a couple of benchmarks for amounts of power and energy:

solar power – The power pouring onto Earth in the form of sunlight (ignoring reflection and averaging over the Earth's surface) is about 340 watts for every square meter of area. Since much of the energy available to us on Earth derives ultimately from this influx of sunlight, keeping this number in mind is handy as a point of comparison for different energy sources and the energy requirements of various appliances and activities. For example, a typical microwave oven uses about 1,000 watts.

total annual human energy use – about 4 x 10^{20} joules or about 10^{14} kWhr. This is a good reference number to keep in mind for thinking about the energy needs of human society compared to the amount of energy conveniently available from various sources.

With these units in hand, we can trace in more detail the kinds of transformations we discussed before. For example, how much food is needed to climb a small mountain? In order for a person with a mass of 50 kg to get to the top of a mountain 1,000 meters tall, we need to take from the food an amount of energy that is given by $E = mgh = 50$ kg x 9.8 m/s^2 x 1,000 m = 490,000 joules or about 120 Calories. A standard candy bar or granola bar should cover that.

Of course this is a bare minimum; we expect to need more than 120 Calories to climb the mountain because our bodies are far from perfectly efficient in transforming food energy into gravitational potential energy. Much of it goes into heat as part of the process of moving us up the mountain. But you get the idea that there is a direct relationship between the amount of food available and the height to which we are able to climb—quantities that at first glance have nothing at all to do with one another. They are linked by energy.

A question may have occurred to you in thinking about all this: Why do people worry so much about "conserving energy" if it is a fundamental law of nature that energy is always conserved? The reason is that not all forms of energy are equally useful for converting into the specific form you want. For example, maybe you need about 2,000 Calories to provide the energy for a typical day of activities. It turns out (by a calculation very similar to what we just did for climbing a mountain) that this is the amount of gravitational potential energy stored in a 100 kg weight held at a height of 10 km above the surface of Earth. So in principle, if you are hungry, a 100 kg weight falling on you from a height of 10 km should appease your hunger and provide you with your daily supply of energy. I don't recommend trying this! Your body has no mechanism for converting the energy in the falling weight into forms that drive the chemical processes that keep you alive. However, you are still better off having the gravitational potential energy available than if you had no source of energy at all. You could for example use the falling weight to turn a generator that powers a light to grow plants which you could then eat; effectively packaging the gravitational potential energy in the form of chemical bonds that your body is able to use.

When we talk about "conserving" energy, what we really mean is keeping it in forms that are *useful* to us for particular functions. Gasoline, for example, is a much more useful form of energy than the heat stored in the random motions of molecules after the gasoline has been used to drive our car around. The energy is all still there, it just isn't in forms that we can use.

As you notice transformations in your life and your surroundings, become aware of the connecting thread of energy that flows through them all. See if you can trace the path of energy as you take a bite of food, turn on an appliance, ride your bicycle up a hill, or drive your car to work.

Chapter 7

Chemical Connections

Take a moment to reflect on what historical figure (someone who died at least a few years ago) you would most like the opportunity to meet. Then take a deep breath and hold it for a moment. You now hold in your lungs many atoms of nitrogen and oxygen that were once in that person's lungs. In fact, you probably hold in your chest an atom that was exhaled in that person's *last* breath.[5] You are connected to that person and a thread of their breath and life now continues through you.

To understand how this can be, we need a few facts. Each breath contains about 1 liter of air, which amounts to about 10^{22} air molecules (mostly nitrogen molecules, N_2). It also happens that Earth's atmosphere contains about 10^{22} liters of air. Thus when fully mixed, each liter of air contains about 1 molecule from any given breath exhaled by any given person!

We can extend the chain of connection beyond our fellow humans. Those nitrogen and oxygen atoms existed in Earth's atmosphere long before humans walked the planet. Even further back in time, as we discussed in chapter 1, the nitrogen and oxygen were themselves formed billions of years ago in the cores of stars, built up from lighter elements.

This process continues today in the core of our own star, the Sun, which is building up heavier elements that will one day form the material of future planets and perhaps life. This insight offers another thread in the flow of transformations that connect us to the universe. Rather than following the path of particular particles such as nitrogen or oxygen, we can trace the flow of *energy* (see chapter 6), which can transform from matter to light to heat, for example.

On a sunny day, let the sunlight fall on your face and hands and arms. Feel the warmth streaming from the Sun. As you're basking in this energy consider this fact: during each second our Sun is losing about four billion kilograms of mass. It's converting hydrogen into helium in its core via nuclear fusion (the process that formed, in other ancient stars, most of the elements that make up your body). You probably weren't too concerned about this, but just in case, know that the Sun has a total mass of about 2×10^{30} kilograms, so you need not worry about the Sun wasting away anytime soon!

This lost mass *is* the energy that is transformed into the light we see and feel. Following Einstein's famous relation $E = mc^2$ (recall chapter 6), mass, light, and the warmth we feel in our bodies are all different kinds of energy and we can trace the flow of energy from one form to another. The light we feel, the warmth in our bodies, even the food we eat owe their existence directly to the decreasing mass of the Sun.

To make this transformation more concrete, consider that four billion kilograms of water (or any material with about the same density as water) would fill a cube-shaped container about 160 meters on a side. So picture in your mind a cube of water (or perhaps chocolate pudding, if you find water too boring) that roughly fills a football stadium. Now let the pudding vanish and replace it with a spherical shell of light 300,000 kilometers thick (the distance light travels in a second) moving outward at the speed of light from the place where the pudding used to be. This is essentially what happens in the Sun every second. It has been happening for over four billion years, and will continue to happen for several billion years more. A little bit of this shell of light is what you feel on your face when you stand outside on a sunny day. Something to ponder the next time you're feeling alone and isolated. The energy of the Sun and other parts of the cosmos is literally within you, flowing through you, each moment.

Chapter 8

Fields of Dreams?

Close your eyes (after reading the rest of this paragraph for instructions!) and try to describe as much as you can about the room you are sitting in. (Or your immediate surroundings, if you happen to be outside.) How big is it? (Or how far can you see?) What's on the walls (or trees)? What objects are present and what color are they, what shape and texture?

It's difficult to take notice even of the obvious things around us, as you'll probably discover when you open your eyes and look around to see what you left out of your description. And there is much more going on beyond the obvious. In focusing on vision, we've left out smells and sounds, for example. And much is beyond detection by direct human senses. In your description, did you mention the free music, videos, credit card numbers and secret love letters that are floating in the air around you, if you only you knew how to read them? Actually, you probably do know how to read some of them—just get out your cell phone or computer to translate into a form you can read.

Before being translated by one of your electronic gadgets, this information lives in something called electromagnetic fields (a word derived from electric and magnetic, the two forces involved.) Let's explore a bit to see if we can get a deeper understanding of all that information flitting about invisibly through the air (and through *you*, for that matter).

You can start getting a feel for magnetic forces by picking up a couple of ordinary refrigerator magnets. Play with their orientations until they repel when you try to push them together. That force you're feeling is the magnetic force, which clearly exerts an "action at a distance." You feel the force even when the magnets themselves are not touching. Remember that point for later.

Electric forces are similar. (In fact, it turns out that electric and magnetic forces are fundamentally expressions of the same underlying force. Magnetic force is produced by the *movement* of electric charges such as electrons.) A good way to explore the electric force is to blow up a balloon and rub it against your hair or a sweater. The rubbing transfers some electrons and leaves the balloon with a slight excess of charge. The balloon will now stick to a wall, or move little bits of paper. Again, the force acts even from a distance. You see the influence even when the balloon is not touching the wall or bit of paper.

Let's sum this up with a mental model that cuts out most of the details. We'll focus on the electric force for this illustration. If you haven't already done so, actually get a balloon, blow it up and give it a negative charge by rubbing it on your hair or shirt (or a cat if you have one handy). Hold the balloon out and stare at it for a moment. This is your negative charge, which we will call the "source charge" for this illustration. All that matters for our model are its location and its electric charge. So you can picture in your mind's eye just a small sphere with a "minus" sign written on it to indicate the negative charge. You can forget about all the other details such as its color, shape, etc. You can test that this source charge exerts a force on other things by bringing it close to a tiny bit of paper, your hair, or your cat if he's foolish enough to still be nearby. For our model, the only thing that matters about these objects is that they also have an electric charge, so we'll refer to them as "test charges." Negative test charges are repelled by your negative source charge, positive test charges are attracted to your source charge. Let's pick a particular test charge, say a positive one, and represent it in our mind as a small sphere with a "plus" sign drawn on it. (Just as with the balloon source charge, all we care about is the location of the test charge and the amount of positive or negative charge it has.)

Imagine drawing a little arrow in space to represent the force exerted by the balloon (source charge) on a bit of paper (test charge). The arrow points in the direction of the force (toward the source charge in this case) and the length of the arrow indicates the strength of the pull. Now imagine moving the test charge around to different locations and drawing the corresponding force arrows. All the arrows will point toward the source charge, and they will get shorter as you move farther away from the source. (Indicating what you already know intuitively, that the force is stronger the closer you are to the source charge.) Leave the arrows and take away the test charge, and you have a *field.* The set of all the imaginary arrows, at every point in space, specifying the force that a test charge *would* feel *if* placed there, is the *electric field.*

This might have just been a handy tool for describing what's going on. But electromagnetic fields turn out to have a life of their own. For example, they are a form of energy in the spirit of our discussion in chapter 6, so they are real in that sense at least. Perhaps the most direct way to understand that electric fields are real is to consider this question: What happens if I suddenly *move* the source charge? (Hold the balloon out to your right and then quickly move it over to your left side.) We can guess that somehow, the arrows will adjust to point toward the new location. But the deeper question is, *"Does this happen instantaneously, or is there a delay?"*

Ponder this question for a moment, before you read on. It's helpful to consider a related question: If you tie one end of a rope to a door knob, walk across the room holding the other end and stretch the rope tight, then quickly move the end of the rope you are holding, will the entire rope adjust instantaneously to where your hand is, or is there a delay?

It turns out that there *is* a delay (in both cases), and guess what? The lag in the electric field case is just the same as the time it takes *light* to travel that distance! *The speed of the kink in the field is the same as the speed of light.* This insight led to the realization that light is a wave in the electromagnetic field, analogous to a wave on a rope as you move your hand back and forth.

So to pull this all together: visible light, along with every other form of electromagnetic radiation (e.g. microwaves, radio, x-rays, etc.) results from wiggling little charges around (like the ones on your balloon) so the field lines they produce are shifted. These kinks in the field lines travel at the speed of light. To make different kinds of radiation, just wiggle the charge at a different frequency. Since the speed of light is very fast (300,000 km/sec), you have to wiggle the charge back and forth very quickly or else you end up with very long wavelength light. For example, shaking your charged balloon back and forth once a second produces radio waves with 300,000 km separating one wave crest from the next. (Think back to the kink on the rope, and imagine the kink traveling away from your hand at 300,000 km/sec as you shake it to make waves.)

This means all that information floating around in space consists of wiggles in the electromagnetic field lines. These field lines represent the forces due to charges in different places, with values (imaginary arrows with length and direction) that exist invisibly at every point in space.

Pause and consider for a moment all the electromagnetic fields that are permeating your life right now, responsible for television, cell phone, wireless internet, cooking your food, etc. Hold out your cell phone and imagine how many conversations are happening right now, passing through the space your phone occupies. As a matter of fact, it just occurred to me that I'm writing this chapter on my iPhone, so my thoughts as I write are traveling through electromagnetic fields to "magically" sync with a server and then with my computer. Consider what that means: This chapter *about* electromagnetic fields is literally stored *in* electromagnetic fields!

Chapter 9

Quantum Reality

There is perhaps no better way to come across as both intelligent and esoteric than to tell someone you're studying "quantum physics."[6] Despite its reputation as an abstract subject, technology resulting from quantum physics comprises roughly a third of our economy[7] and the shift in worldview it suggests is perhaps even more important than its economic value. Quantum physics has implications for our most basic notions of reality and what it means to experience and know something about the world. If nothing else, it's worth learning the basic insights of quantum physics simply as a reminder that the world is often not what we think it is. Once you have really absorbed this shift, it leaves you more open-minded about everything.

There's a basic question to keep in mind as we ponder the conceptual lessons quantum physics has to teach us. It has occupied the minds of Einstein and many other thinkers since the development of quantum theory in the 1920s, and is arguably *the* core question: Do physical properties (like position, speed, direction, mass, color, etc.) have definite values independent of any observation of them? As Einstein once supposedly expressed it with much frustration over the implications of quantum theory, "Do you really think the Moon is only there if you look at it?"

At first this seems like an unanswerable question. What testable difference can it make whether the Moon is only there when you look (springing into existence when you observe and going away when you don't), or whether it's always out there, maintaining its properties independently of whether anyone or anything is looking. As long as it's there whenever you do look, how could you tell the difference? But as it turns out, it does have testable consequences. If you insist on believing properties are "there" in an absolute sense when nothing is "looking," you'll make predictions that are inconsistent with what is observed.

Physicist John Wheeler liked to illustrate the shift quantum physics suggests in our way of viewing the world with a "surprise" version of the standard game of twenty questions. The standard version goes like this: One person thinks of an object and other people ask questions to evoke yes or no answers, trying to zero in on what the object is. If they guess it before getting to twenty questions, they win. If they use up their twenty questions without guessing, they lose. In the analogy, the person thinking of the object plays of the role of nature and the guessers play the role of the experimenters.

In the surprise version, the person playing the role of nature does *not* think of a definite object ahead of time. They just have to be consistent in how they answer questions. (They can't say "yes" to both "Is it smaller than my hand?" and "Is it larger than Earth?") The experimenters ask questions and receive answers just as before. But notice a couple of profound differences in this version of the game:

1) The questioners help determine the outcome by what questions they ask and in what order.

2) Play the game again with the exact same questions, and the result may be *different*. It's a dance—an interplay between questioner and answerer, experimenter and nature.

The world as described by quantum theory acts much like the surprise version of twenty questions. By no means is it a free-for-all. There are definite rules and it doesn't mean you create reality completely by your own will or whim (as is sometimes implied in popular literature or movies). It's just a bit richer and more flexible.

A nice way to get a feel for how it can make a noticeable difference whether the answer to an experimental question has an independent reality is to consider a pen oriented on a sheet of graph paper described by x (left-right direction) and y (up-down direction) coordinates. If the pen has a definite orientation on the paper (an "absolute reality"), then the components of its length are restricted by the Pythagorean Theorem: $L^2 = x^2 + y^2$. That is, if you take a real pen, measure its length L with a ruler and then also measure how much of that length is in each of the x and y directions, the fact that it is a "real" pen with a definite reality in space *requires* that these three measurements be related by the Pythagorean Theorem. But without this requirement there's more freedom for the components, when measured, to have different relationships. They still have relationships, just more freedom to be something other than the Pythagorean Theorem. Just like in the surprise version of twenty questions. The answers must be consistent but are not bound by being different aspects of a single specific real object.

A variety of experiments in quantum physics can be viewed as setting up situations where relationships analogous to the Pythagorean Theorem are violated in the real world, in ways that match the predictions of quantum theory.[8] We'll consider one example to illustrate the essential features.

Consider two particles which have interacted and then traveled far apart. Because of their interaction, they are correlated or *entangled*. (The entangled pair of particles is often referred to as an EPR pair in honor of a famous 1935 paper by Einstein, Podolsky, and Rosen which motivated these experiments decades later.) In real experiments, the particles might be photons or electrons and the entangled properties might be the polarization of the photons or the spin direction of the electrons. For simplicity in this cartoon illustration we can think of the particles simply as arrows correlated by the requirement that when you measure their orientations, they must always be the *same*. (If one arrow is up, the other will also be up, etc.)

The experiment amounts to creating entangled EPR pairs of particles, letting them travel apart from each other, and then measuring the orientations of the arrows relative to some coordinate system (like the horizontal and vertical lines on a piece of graph paper). This experiment is repeated many times, with the coordinate system rotated at different angles. There are many technical points to consider in the real experiments. But the essential features (expressed in our cartoon version) are these:

1) The orientations of the arrows are always measured to be the *same*. (As we mentioned before: when one arrow is up relative to a particular coordinate system, so is the other arrow.)

2) The orientations of the arrows are *random*. You can't predict ahead of time whether the arrows in a given experiment will be up or down.

3) By considering the results from many different runs of the experiment, you can show that the arrows could *not* have possessed well-defined orientations prior to being measured. The argument is analogous to our earlier example of showing that the components of the length of an arrow do not satisfy the Pythagorean Theorem.

You can get a feel for the strangeness of these results by considering an even more cartoon version that captures the essential features. Suppose you and I meet up and agree that we will go off to opposite corners of the country and write down the first twenty numbers (restricted to 0 or 1) that pop randomly into our heads. Your list might be 00101110100110000011. No big deal—just a string of random binary digits. But imagine your surprise if we met up again later to compare lists and found that mine was *identical* to yours, also reading 00101110100110000011. This is essentially the situation in the quantum experiments. It seems that the particles are somehow communicating instantaneously across a distance, to ensure that they both point in the same direction every time!

Pause a moment to ponder the implications. It's often said that the quantum world is very strange and counterintuitive. But it's only strange in comparison to our *expectations* about how the world is supposed to be. These expectations in turn are products of the particular model we hold in mind. We think the quantum world is bizarre because we are wrapped up in a limited mental model of the world which assumes everything is made of a kind of "material stuff" which has particular properties that pervade our thinking and guide our expectations. It's challenging to let go of this mental model, but well worth the effort.

If we really pay attention to what nature is telling us and let go of our preconceived material stuff model, the experiments are much simpler to interpret. For example, as you read about the entangled EPR pairs whose orientations always came out the same, didn't you picture two arrows separated by some distance, and then wonder what kind of signal was sent instantaneously from one to the other to make them line up? But this mental image is almost certainly wrong. It's an artifact of the stuff model, which makes you think there must be arrows already in place with definite locations, waiting to have their orientations determined. The paradox diminishes if you recognize that the locations and orientations are *both* answers to questions about nature, constrained by past observations and the laws of physics. Other than the assumption imposed by our stuff model, there's no reason to think the arrows even existed as arrows until they were measured. It seems that the twenty questions game is an even more accurate description of the situation than we might have thought. Nature gives consistent answers to questions that are asked, but is not required to fill in the gaps with definite objects waiting in the background to answer questions that have not been asked.

Without the interference of the material stuff model, the wave-particle duality you've probably heard about as a basic feature of quantum physics is also much easier to swallow. For example, you may have heard that an electron sometimes behaves as a wave and sometimes as a particle, and wondered how that could be. You're probably picturing a little electron that looks something like a marble (maybe with a small "minus" sign etched on it), and wondering how this solid object with a definite location could dissolve into a spread out wave sometimes, and then condense back into a localized particle at other times. One sensible answer, in the spirit of our twenty questions analogy, is that the electron is never either a particle or a wave. Those ideas are artifacts of our stuff model. An electron is simply an expression of certain restrictions nature imposes on the kinds of answers that can be given to questions that are asked.

In a way, this lesson from quantum physics just continues the evolution in our thinking we started with the discussion of energy in chapter 6, fields in chapter 8, and a little bit with our discussion of time and relativity in chapter 4. All of these developments lead us away from the idea of a world made of solid stuff that has an absolute reality, and more toward a view of the world as described by the surprise twenty questions game.

Although this new way of looking at the world is suggested by quantum experiments, it may help resolve other puzzles in our understanding of the world. For example, the mind-body problem or the so-called "hard problem of consciousness" seems almost intractable from within the stuff model.[9] (How can it be that the material brain gives rise to the non-material mind and to the experience of awareness—that it feels like something to be you or me?)

I think the experiments of quantum physics are best viewed as a reminder that whatever model of reality you are carrying around in your mind right now, it is never the whole story. Our mental models shape our perception of reality. See if you can notice each day an instance or two where you assume something based on a view that the world "really" is made of material stuff that forms the building blocks out of which everything else is made. And then notice what new possibilities open up in your mind as you relax the grip of a particular model on your way of seeing. These moments that occur when you step a little bit outside of your current model are moments of magic. As John Welwood writes, "Magic…is a sudden opening of the mind to the wonder of existence…a sense that there is much more to life than we usually recognize."[10]

Chapter 10

Infinite Possibilities

Infinity is a mind-stretching concept even as a mathematical abstraction. For example, consider mathematician David Hilbert's imaginary hotel. Hotel Infinity boasts unlimited rooms, numbered 1, 2, 3, and so on. Business is good, and the hotel is completely full—no matter what room number you pick, that room is occupied.

A weary traveler arrives at the front desk, hoping to get a room for the night even though the sign outside says "no vacancy." The desk clerk is wise in the ways of mathematics and eager to make as much money as possible. "Sure, no problem," she says. "Just give me a few moments and then you can have room #1." She makes a quick call that is broadcast to every room in the hotel, asking each occupant to move to the room numbered one higher than their current room. (The key cards also have wireless receivers so she can automatically switch each key to open the room numbered one higher.) Thus the former occupant of room 1 is now in room 2, room 2's guest has moved to 3, and in general, the person in room N is now in room N + 1. She then collects the weary traveler's credit card and hands him a key for the now-empty room #1 in return. Everyone is happy.

If you're puzzled by this you're not alone. What's even more puzzling is that the same process could be repeated to make room for as many additional guests as you care to name— even though the hotel was already filled! The key to understanding how it's possible is to recognize that infinity is *not* a number. Rather, it's a label for the condition of being *unbounded*. With a finite number of rooms, moving everybody to a higher room number leaves someone out in the cold. You could identify the "last" room in the hotel and see that its occupant has been kicked out, with no extra room to move into. But in an unbounded hotel this doesn't happen. There is *no* highest numbered room.

Now that we've stretched our minds with the concept of infinity in an imaginary situation, let's return to our physical universe. Astronomical evidence points to a very real possibility that the universe may be unbounded, like Hotel Infinity. What's more, evidence suggests that the average density of matter on large scales remains uniform throughout this unbounded space. That is, wherever you are in this infinite universe, from a big picture perspective your surroundings look roughly the same as they do here. If you were suddenly transported to a location gazillions of light years from here, you'd likely still see galaxies spread through space in a clumpy, filamentary pattern, each galaxy a few million light years from its neighbors on average, each galaxy containing a hundred billion stars or more on average, with many of these stars orbited by planets, etc. Just as we observe in the space near us.

Pause a moment to let this sink in and begin to consider the consequences. In an infinite space filled with matter it seems unavoidable that every possible arrangement of matter that can happen (no matter how tiny the probability) actually does happen somewhere in the universe. And not only does it happen, but it happens an unlimited number of times!

This suggests that somewhere far across the universe there must be another you, identical in every respect to the "real" you that we know and love, reading this page. And also someone similar to you but doing something other than reading this book. And one just like you but with green and purple hair (or with brown hair, if yours is green and purple). And so on. Even farther away there are still other repeated versions of you with each of these variations, and so on without bound.

The mind reels at the implications. Think of every story you've ever read that does not violate the known laws of physics. Somewhere in the universe (perhaps infinitely many times) that story may have actually played out just exactly as in the fictitious story you read![11] Somewhere a version of you may live on a planet located within a globular cluster so the sky is filled with stars. That planet may have a thin atmosphere (just enough to enable you to breath) and low gravity so there are sentient beings with bodies of gossamer fabric, very light and thin and shimmery, forming fronds that spread out like a canopy that flutters in the breeze. They're extraordinarily kind and loving and when you lie on the ground (which is comfortably soft and spongy of course) looking up at the breathtaking stars, they gently caress your cheeks and arms so you feel surrounded by soft, warm, loving energy. Somewhere far across the universe, is that scenario actually playing out right now? Is another you, identical to yourself, savoring that experience?

This whole idea is both reassuring and disturbing on multiple levels. On one hand it's quite comforting to imagine, as you agonize over some life- or civilization- changing decision ("paper or plastic, or perhaps a fabric reusable bag?"), that no matter what you decide, all other possible decisions may also be playing out in other parts of the universe.

On the other hand, in order for it to really matter what we individually choose to do, it seems that there must be some outcomes that never happen anywhere. If inevitably on some earths human civilization survives into the distant future while on others we destroy ourselves within a few decades, then what possible difference can it make which category our own Earth falls into? Either case will happen an unlimited number of times no matter what we do. It seems as if it only matters what we do if there is some fundamental uniqueness to us and our situation.

This gets to the heart of the matter: It's very disturbing to our sense of identity. Does each of your distant copies actually experience itself as you, with the same feelings and the same sense of being the "real" you, as you experience? Is there a sort of "generalized Copernican principle" (no privileged point of view) according to which none of the versions of you has any more claim than the others to being the "real" you? If so, then where does our sense of a unique identity come from?

Perhaps more than any other example this reflection illustrates the power of truly con-

necting everyday human experience to a cosmic perspective. On the one hand, nothing
seems more esoteric, abstract, and distant than
the question of whether there is another you
identical to yourself, across the universe so far
away you could never possibly meet. And yet
on the other hand, nothing is more directly rel-
evant or strikes deeper into the heart of your
everyday experience. Can you really take the infinite universe possibility seriously as you agonize about your situation in some immediate
moment? When you are crippled with regret,
haunted by a missed opportunity, sure that you
made the wrong choice, can you take consola-
tion from the possibility that all other choices
happened as well, elsewhere in the universe?
(We'll explore this notion of counterfactuals,
the "what might have beens" a little further in
chapter 12.) If you seriously see every possibility happening somewhere, and all those as
equal (the one you're aware of no more special
than others), then it should greatly impact how
you feel about that situation. Can you connect
it concretely so it does? If you really believed

this idea at a gut level would you ever feel hesitation about making a difficult decision? Once you've made a decision could you really be that upset about an outcome that wasn't what you wanted in that case?

There may very well be a flaw in the argument that everything which can happen does happen somewhere.[12] There's something fishy about the assumption that an identical physical copy of you would actually *feel* exactly like you. But it's also hard to pinpoint exactly what else there is to adjust about the copy in order to make it feel different than you.[13] And maybe the universe simply is not infinite.[14] But one thing is clear: Either the universe is infinite or it is not. In either case the implications are profound. The answer dramatically impacts our understanding of what's really happening, and why it matters, when we make the simplest of everyday choices.

Chapter 11

Thinking About Nothing

Compared to the wondrous expanse of infinity, pondering "nothing" seems a very boring prospect. In fact, isn't nothing the epitome of boring since by definition there can be nothing interesting going on where there is nothing at all going on? But since we have nothing to lose... well, let's spend a few moments reflecting on the concept anyway.[15]

First just try to picture nothing in your mind's eye. Imagine taking away all the obvious physical objects in the room: yourself of course, and your desk, chairs, books, pillows, iPod (even simple quietness is hard to fathom, much less true nothingness!).

Now take away the things you don't see but know are present: air molecules; dust; a tiny spider hiding in the corner of the ceiling.

Next remove the things you may not even know about: cosmic rays and neutrinos from space that are streaming unnoticed through your body right now; electromagnetic radiation of all sorts (recall chapter 6) including all colors of visible light, infrared and UV light, wireless internet, your cell phone signal, and radio waves from around the universe; gravity waves; other forms of energy as yet unknown to human awareness.

Are you starting to feel pretty good about your progress toward nothingness? You're picturing a dark, empty space, very uniform, with nothing to break up the monotony of emptiness. But we still have a long way to go to reach true nothingness.

Try taking away the empty space that probably represents your first approximation to nothing. Even the fact of space having three dimensions (or however many it actually has) is certainly *something*. As evidence of this, note that our lives would be very different if we lived in a two-dimensional world, a "flatland."

Time must go away as well. And your thoughts and feelings, if those didn't already vanish in your mind's eye when you took away your body. This means we must remove from existence the very capacity we're using to try and *imagine* nothingness. Oh, and the laws of nature themselves must be banished: the one that makes sure energy is conserved and the one that sets the strength of gravity and the speed of light. Even the one which ensures that after repeated washings and dryings you will be left with an odd number of socks. Not feeling so great about your progress toward nothingness now, are you?

It almost seems impossible to ever fully comprehend the notion of nothing. This is apparent in the age-old struggle to understand how "something" can arise from "nothing." However we twist our minds to try and resolve this dilemma of origins, we always end up with an image of our present cosmos arising from something *else* that is not completely nothing. Anytime we try to imagine something coming from nothing, the starting point is a fertile nothing that is not really nothing—e.g. it contains already the laws of quantum physics, time, space, etc.

This same truth is more succinctly captured by the Australian Aboriginal saying, "nothing is nothing." [16]

We are left with the feeling that "nothing" is impossible. Even if we think we have imagined nothing, we are doing so within *time*. And there could never have been a time when nothing existed, because time itself is something. It seems we are stuck in an impenetrable paradox. So why did we put ourselves through all these words and mental gymnastics only to return, apparently, to the same state of bewilderment from which we began? Because it sheds new light on the paradoxes that abound in everyday life. It's good to ponder questions (and even take actions) that make no sense once in awhile, if only as a reminder that existence itself makes no sense in any conventional way of thinking.

We get stuck sometimes in the paradoxes of life. But when we remember that existence itself is a paradox at its core, the puzzles of everyday life seem less troublesome. The next time you're feeling exasperated by your friend's behavior, remember that his very *existence* makes no sense. Everything, including your friend and his crazy behavior, is frozen creation. It embodies a tiny piece of exactly the mystery we've been wrestling with—why anything exists at all. So take a deep breath and then relax, let it be, and savor the mystery.

Chapter 12

Counterfactual Squirrels

"'There is no use trying,' said Alice; 'one can't believe impossible things.' 'I dare say you haven't had much practice,' said the Queen. 'When I was your age, I always did it for half an hour a day. Why, sometimes I've believed as many as six impossible things before breakfast.'"

~ Lewis Carroll

I was riding in the car with a friend the other day and watched as he swerved to avoid a dead squirrel in the middle of the road. My initial reaction was the expected, "Oh, poor squirrel!" But then I began to wonder. Why do I respond emotionally to that particular poor little critter and not to all of the *potential* squirrels that might have existed but don't? For the rest of the drive I found myself peering along the roadside as if looking at all the squirrels that *might* have been sitting in each open space on the curb, nibbling happily on acorns. But they don't exist (at least not in this part of the universe—recall chapter 10). Why doesn't *that* feel sad to me? We react to things that once existed and then stop existing, but not to things that never existed at all. Why don't we experience an emotional response to counterfactual squirrels?

On one level the answer is pretty obvious: Our ancestors would not have survived very long if they worried much about such hypothetical things. *Factual* squirrels (or saber-toothed tigers) are of far greater concern than counterfactual ones. Thus our brains have not evolved to worry about them. In fact this reflection on counterfactual squirrels began simply as a joke. But the more I pondered it, the more I realized that this line of thought has much to teach us about the nature of reality and what matters. Everything meaningful in our lives happens at the interface between the real and the counterfactual. The interplay between two levels, two worlds—reality and imagination, factuals and counterfactuals—is at the core of what life is all about. (Plus it seems appropriate to follow the discussion of "nothing" with a discussion of "something," but something that does not actually exist!)

Consider anything you feel concerned about that spins around and agitates in your mind. For example:

"I should have held onto that Apple stock I bought ten years ago at $22 a share!"

"Why didn't I talk to that cute guy or girl I saw on the bus?"

"If that wide receiver had just held onto the ball in the end zone, my team would have won the game."

"If I had been born in an era before penicillin, I would not be alive today."

"If I had kept my cool and not become angry or upset, my friend would not have abandoned me."

"If I had chosen that other school or job, I'd be happy right now instead of miserable."

All of these examples depend on the notion that there is a way things *are*, and also a way things *"might have been."* The "woulda, coulda, shouldas" of life. Our belief that we even have choices to make relies on the existence of counterfactuals. What happens in the transition from counterfactual to real? Do the possibilities somehow "already exist" out there and we move from one to the other depending on our choice? Or do they come into existence with our choice, so that we create them…out of what? What exactly happens during that moment of making a choice? What power do we have, or is the sensation of making a choice an illusion somehow? It's worth noting the connection to our earlier discussion of infinity (chapter 10). Is everything that can possibly happen actually happening, and somehow our awareness moves around to these different worlds depending on our choices?

Flying on Southwest Airlines (with no assigned seats) provides a microcosm for exploring counterfactuals. Unless you check in online during a very brief window lasting about thirty seconds (between 24 hours before your flight and 23 hours, 59 minutes, and 30 seconds before the flight), you most likely won't be in the coveted "A" boarding group. Which means you board the plane with most aisle and window seats already taken. But unless the plane is very full, you likely still have many options among middle seats. What kind of conversation might you have sitting next to this person, or that one?

If you really stopped to think about it, the choices would be overwhelming even just among one planeful of people. And this is only a tiny fraction of all the people you cross paths with during your lifetime. Might you have made a lifelong friend or found your soulmate if you had just chosen the seat one more row farther back? What dramatic change might have occurred to your life if only there had been a little more room in the overhead bin above seat 23 B and you had settled there instead of all the way back in 27 B?

Thankfully, we don't normally think too much about all the choices before us. We just pick a seat and enjoy (or grumble internally about) whatever conversation we end up having as a result. But in other areas we *do* feel the full impact of the might have beens. If you've just broken up with a boyfriend for example, your mind is quite possibly flooded with "might have beens" you wish you could turn off to give yourself peace. At one level, your former boyfriend is not fundamentally different than the hundreds of people you might have gotten to know on the flight from Portland to Kansas City. But on another level, the closeness makes him feel irreplaceable to you in a way that hundreds of strangers do not. So you're not upset about the "loss" of these "counterfactual relationships" the way you are about the loss of a factual relationship.

This brings us back to the poor squirrel (the one on the road, not your ex-boyfriend). We're bothered most by counterfactuals that are "closest" to reality in some way. The dead squirrel on the road pulls at our heart strings because it is so *close* to being alive still. If only it had darted a couple of inches more to the left, it would have missed the tire and been safe. In fact not long ago it *was* alive. By contrast, the imaginary squirrels never were alive to us, so we don't feel the loss of what might have been. The cute guy or girl on the plane doesn't throw us into an emotional tailspin the way an ex-sweetie does because we don't feel the tug of the deep sense of loss, of what once was but now is not.

Along a similar line, ask yourself which of the following scenarios you find more upsetting:

1) If you lived in a larger city, it might have been home to a professional basketball team which might have made it to the NBA playoffs last year and won the championship.

2) You *do* live in a large city with an NBA team which you follow closely and they lost on a 3-point shot at the buzzer in game 7 of the championship series.

The second scenario is probably more painful because the sense of loss is greater. *So* close, but not quite. Only a small change, a few centimeters more to the right and the shot would have bounced harmlessly off the rim and your team would have won the championship. In scenario 2 your desired outcome is just as counterfactual compared to reality as in scenario 1, but somehow it feels much more like it "could have been," so it hurts more. And you're much more likely to spend time replaying the game, thinking, "If only this or that had happened just a bit differently, my team would have won and all would have been right with the world." But is that belief really true? Maybe outcomes can only be classified as those that do happen and those that do not happen. Is there a meaningful way to measure the "distance" of any given counterfactual away from reality? And if so, is there a sense in which if it's "close enough" to reality, then it truly could have been real, with a change that was within your power (or the power of your favorite NBA player)?

All of which returns us to the core question of what happens when we make a choice. Where do counterfactuals "live" that is distinct from where reality lives, and how is it that we can turn one into the other? The question takes on even more urgent force when it comes to directly personal choices over which you feel you have some immediate control. (Yes, I know, your team's loss in the NBA final was personally devastating to you, but still there's a difference...) Your path through life has so many branch points, so many places where a minor change in your choices (or other tiny events beyond your control) could take you along vastly different branches. Are you in any real way responsible for which branch you follow? Is there a sense in which you "should have known" not to say the wrong thing to your girlfriend, and with the "correct" course of action you'd still be together? If you are diligent enough about eating and exercising in the right ways, will you stay healthy? Is there a set of actions I could take, and should be able to discern, that will result in me discovering the cure for cancer? Or on a more global level of social change, what sense are we to make of the idea of progress? Moving from a counterfactual, imagined state of how things should be (more fairness, less hunger, longer and healthier lives) to create a real world that more closely matches that hoped-for state. This idea is the basis of morality—a sense that there is a way we "should" do things that we can recognize if we are vigilant enough, but that we are capable of choosing to do or not do.

And in all these cases, how do we know that the different state of affairs we might imagine or hope for is actually preferable? How often have you believed something would bring you eternal peace or happiness (anticipating a perfect job or relationship or ice cream sundae, or your team winning the NBA championship), only to find that the glow soon fades and you find something else to be dissatisfied about? Counterfactuals have a magic to them and also an incompleteness, that reality does not. We're allowed to leave gaps within counterfactuals that don't exist within reality. We forget, while in counterfactual-land dreaming about how glorious it would have been if that last second shot had been blocked, that afterward we still would have to get up the next day, deal with traffic, our boring job, etc. In the counterfactual world all we see is the moment of winning the game or the intensity of that perfect kiss that makes your heart skip a beat.

We also forget what else might have to change in order to get the one piece we are focused on at the moment. The present state of affairs is probably pretty good compared to all the things that might have been. You're worried about whether you could have made a little more money in Apple stock. But what about all the worlds in which you were born into slavery, or starvation, or didn't exist at all? We think we can make changes and take only the good, ignoring the undesirable outcomes that might be linked to the same small changes.[17]

Counterfactual world has lower resolution than the real world. It leaves out the details. This allows us to assign a certain specialness and magic to counterfactuals that are denied to reality. What if we could learn to see this as a matter of perception? After all, today's counterfactuals are tomorrow's realities. (And for that matter, yesterday's realities are today's counterfactuals, as we tend to see the past through a distorted lens as well. Only the present is fully real.) Why can't we assign each moment of existence the magic of a counterfactual from that perspective? Then perhaps we could be at peace with experiencing the specialness of exactly what *is*. Can you feel that directly every time you feel stressed about an important decision? The awareness of the creative power you are expressing, bringing something out of the mists of the counterfactual into the concreteness of the factual? And can you recognize that this creative power is as awesome, wonderful, and mysterious as the creative power that makes it possible for anything to exist at all?

Who would have guessed that even roadkill can be a glimpse of wonder, a portal into the deep mystery of existence. May he rest in peace!

Chapter 13

What's Next?

In writing this book, I've picked out a few particular glimpses of wonder to highlight. But the point is that such portals into the deep mystery are *everywhere* (even taking the form of dead squirrels on the road). The trick is in learning to see them and bring them to your own awareness. Anytime you feel stuck, try looking at the world with "beginner's mind"—the perspective of a young child or an alien seeing your world, your civilization, for the first time. What can you notice from this perspective of open receptiveness? What seemingly mundane parts of your life become deep mysteries to explore?

For example, this morning I was making orange juice. Boring. You know, the kind that comes frozen in a can and you forget to thaw it out in advance. So you run warm water over the outside to loosen the frozen cylinder of juice and then squeeze it out of the can and into the pitcher for chopping up and mixing with water. I was doing this absentmindedly until I hit a snag. I was using a new pitcher, one with a narrow mouth. For some reason the cylinder of frozen juice would *not* fall into the pitcher. I checked that it was melted enough on the edges and would indeed slide out of the can. Just not into the pitcher. I tried again. Still no luck. Persisting, I noticed that the resistance had a bit of bounce to it, like pushing against a very stiff trampoline. Suddenly the light bulb came on in my mind along with a little sheepish embarrassment as a trained physicist. *The "empty" pitcher was not empty at all.* It was filled with *air*! The juice can was just the right size to create a nice seal with the mouth of the pitcher so that no air could escape as I tried to cram the frozen cylinder of juice inside. Of course I knew this but it hadn't occurred to me at first, so it was a shock when I realized something that abstractly I knew very well all along: Invisible air is real and at sea level exerts a very substantial pressure. A glimpse of wonder.

What else can you notice in the frustrations of your day-to-day life that transports you into the deep mystery behind things? Try to find and create your own portals beyond the veil in everything you experience. Whatever is happening to you in any moment, how can you connect it to a big picture, cosmic perspective? After all, among the cosmic wonders of distant galaxies and alien life and quantum physics, nothing is more mysterious than the simple fact that anything at all exists in this moment. What will you create with that moment?

Notes

1 This "spider web" metaphor for consciousness was brought to my attention by the writing of Colin Wilson.

2 Joseph Campbell, *The Power of Myth* (New York: Anchor Books, 1991), p. 1.

3 A light year is just the distance light travels in a year, about 10 trillion (10^{13}) kilometers.

4 More on this "time machine" aspect of light in chapter 5.

5 This insight was first brought to my attention by A. Truman Schwartz, and is discussed in more detail by Art Hobson in *Physics: Concepts and Connections*, 4th Edition (New Jersey: Pearson Prentice Hall, 2007), p. 40.

6 See, for example, Jim Carrey's humorous conversation with Conan O'Brien at http://nanobot. blogspot.com/2009/02/jim-carrey-and-conan-obrien-talk.html.

7 Bruce Rosenblum and Fred Kuttner, *Quantum Enigma* (New York: Oxford University Press, 2006), p. 82.

8 There are many excellent resources for learning about these experiments in more detail. At a conceptual

level, a good starting point is Bruce Rosenblum and Fred Kuttner's *Quantum Enigma* (New York: Oxford University Press, 2006). At a more technical level, a great introduction to some of the key experiments is *The Quantum Challenge*, Second Edition, by George Greenstein and Arthur Zajonc (Boston: Jones and Bartlett, 2005).

9 For more details on this, see my article "Untangling the Hard Problem of Consciousness," (*Global Spiral*, November 2009, http://www.metanexus.net/magazine/tabid/68/id/10928/Default.aspx).

10 John Welwood, *Ordinary Magic: Everyday Life as Spiritual Path*, (Boston: Shambhala, 1992), p. xiii.

11 The actual situation in the universe may thus be similar to Jorge Luis Borges' imaginary Library of Babel.

12 Max Tegmark provides a good discussion of the argument in more detail: http://space.mit.edu/home/tegmark/PDF/multiverse_sciam.pdf (*Scientific American*, May 2003, p. 41).

13 There's a deep question lurking here of whether completely describing the physical state of a conscious being actually describes everything there is to know about that being.

14 There are other possibilities consistent with current observations. For example, space could be flat

but "multiply connected," like a video game screen where you fly off one edge and reappear back at the opposite edge. But data from Cosmic Microwave Background observations and galaxy surveys seem to be pointing more and more toward a "flat" and infinite universe. In addition, there are other versions of "multiverses" even if our own universe is not spatially infinite. See Max Tegmark's article http://space.mit.edu/home/tegmark/PDF/multiverse_sciam.pdf (*Scientific American*, May 2003, p. 41).

15 Technically there's a grammatical problem here with talking about "nothing" as if it is a thing, a "something." Rather than using the word "nothing" we should really talk about "the absence of anything." But that's cumbersome, so we'll keep this caveat in mind but continue to use the word nothing to direct our thinking. If this bothers you, please read "nothing" as shorthand for "not something."

16 As quoted by David Christian in *Maps of Time* (Berkeley: University of California Press, 2005), p. 17.

17 Ursula LeGuin explores this sort of idea in *The Lathe of Heaven* (New York: HarperCollins, 2000).

About the Author

Todd Duncan combines a research background in physics and astronomy with experience teaching science concepts to a wide range of audiences. He received his undergraduate degree in physics from the University of Illinois, an M. Phil. from Cambridge University as a Churchill Scholar, and a doctorate in astrophysics from the University of Chicago where he was an NSF Fellow. He joined the faculty of the Center for Science Education at Portland State University in 1997 to pursue an interest in interdisciplinary "big questions" research and its application to science education. In 1998 he founded the Science Integration Institute as a forum for anyone exploring what it means to be human in the universe as understood by modern science. He's the author of *An Ordinary World: The Role of Science in Your Search for Personal Meaning* and co-author of the undergraduate textbook *Your Cosmic Context: An Introduction to Modern Cosmology*. Todd is currently director of the Science Integration Institute and teaches astronomy at Portland Community College.